AERONAUTICAL ENGINEERING ~~~~~~ ~~~~

THE DEVELOPMENT OF SHEET METAL DETAIL FITTINGS

FOR BENCH FITTERS AND TOOLMAKERS, ETC.

BY

WILLIAM S. B. TOWNSEND

SECOND AND REVISED EDITION

First published by Sir Isaac Pitman & Sons, Ltd.
First Edition 1937
Second and Revised Edition 1938
Reprinted 1939

This printing © 2010 Camden Miniature Steam Services

Our thanks to Tim Simpkins for his help in this new printing.

British Library Cataloguing-in-Publication-Data:
a catalogue record of this book is held by the British Library.

ISBN No. 978-0-9564073-4-4

Published in Great Britain by:
CAMDEN MINIATURE STEAM SERVICES
Barrow Farm, Rode, Frome, Somerset. BA11 6PS
www.camdenmin.co.uk

Camden stock one of the widest selections of fine transportation,
engineering and other books; contact them at the above address for
a copy of their latest free Booklist.

Printed by Butler, Tanner and Dennis, Frome

PREFACE

IN the course of a wide experience in aircraft work, and particularly aircraft sheet metal detail work, the author has come into contact with surprisingly few men who can develop their work accurately.

The reason may be that this is a comparatively new branch of engineering, and that no book hitherto has dealt with the subject of developing adequately.

Many men understand thoroughly geometry and trigonometry, but few know how it is applied to the development of sheet metal fittings.

This booklet not only does this, but also caters for the man who has either forgotten and needs revision, or one who has never heard of trigonometry, etc.

The author believes the average bench fitter will have little or no difficulty in understanding the text, as he has had the opportunity of trying it out.

In presenting this book, therefore, the author is not unhopeful that it will supply a long-felt want, and be the means of eliminating wasteful and unsatisfactory "rule of thumb" methods, which are now in widespread use.

THE DEVELOPMENT OF SHEET METAL DETAIL FITTINGS

CHAPTER I

The Circumference of a Circle

THE circumference of a circle $= 2\pi r$, or stated simply $2 \times 3\cdot1416 \times$ the radius of the circle. The symbol π (pronounced "pi") denotes the amount $3\cdot1416$. Take the circle in Fig. 1 as an example, the radius of which is $1\frac{1}{2}$ in.

The circumference

$$= 2\pi r$$
$$= 2 \times 3\cdot1416 \times 1\cdot5 \text{ in.}$$
$$= 9\cdot425 \text{ in.}$$

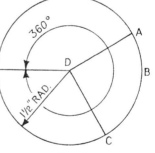

FIG. 1

The Length of an Arc

To find the length of the arc ABC.

There is an angle of 360° at the centre of a circle (Fig. 1). Arc ABC is contained by the angle ADC at the centre. Suppose the angle ADC is 90°, which is exactly one-quarter of the complete angle at the centre—360°. Then, because the angle which contains the arc is a quarter of the whole, it therefore follows that the arc ABC is a quarter of the whole circumference.

The length of the arc ABC is therefore

$$\frac{\text{Circumference}}{4} = \frac{9\cdot425}{4} = 2\cdot356 \text{ in.}$$

The length of ANY arc is $2\pi r \times \dfrac{\text{angle containing arc}}{360}$

1

The Development of a Simple Angle Plate

When a piece of metal is bent, the inner surface of the bend contracts, or *squashes up*, and the outer surface expands, or *stretches*.

For developing purposes the centre line or middle of the metal is reckoned to remain at a constant length, that is, its length does not change during bending. All calculations for amounts of metal used up in bends are therefore made to this centre line. The amount of metal used up in a bend is known as a *Bending Allowance*.

When developing, first make a rough cross-sectional sketch of the job as shown in Fig. 2B.

Commence by separating the bend from the flats, and proceed to calculate the length of each part separately.

From radius point x drop perpendiculars on to the flat portions of the section, shown in sketch as x to B and x to A.

Then the portion M is flat

,,   ,,   N is flat

and length of arc B.A. is the complete bending allowance.

1. Length of portion $M = D - F$.

D would be given by the draughtsman.

The amount F is made up of the inside radius of the bend plus the thickness of the metal.

2. Similarly length of portion $N = C - F$.

3. The bending allowance B.A. is calculated as follows—
Referring to Fig. 1.

$$\text{Length of an arc} = 2\pi r \times \frac{\text{angle containing arc}}{360}$$

Therefore length of B.A.

$$= 2 \times \pi \times \text{radius to middle of metal} \times \frac{\text{angle of bend}}{360}$$

Radius to middle of metal is x to y

= inside radius plus half thickness of metal.

Angle of bend is $90°$

$$\therefore \qquad \text{B.A.} = 2 \times 3 \!\cdot\! 1416 \times xy \times \frac{90}{360}$$

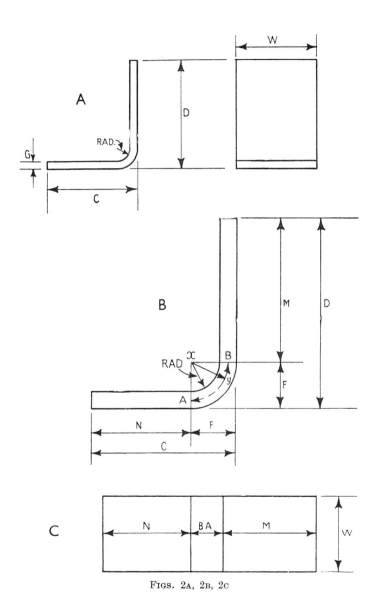

FIGS. 2A, 2B, 2C

The development of this angle plate would be as shown in Fig. 2c.

Example (see Fig. 3A). Rough cross-sectional sketch shown in Fig. 3B.

$$M = 1 \text{ in.} - F.$$

F = inside radius + thickness of metal = $0.125 + 0.064$

$$= .189.$$

$\therefore$ $\qquad\qquad M = 1 - 0.189 \text{ in.}$

$$= 0.811 \text{ in.}$$

Similarly $\qquad N = 1.5 - 0.189$

$$= 1.311.$$

The bending allowance B.A. would be

$$2 \times 3.1416 \times \text{radius to middle of metal} \times \frac{\text{angle of bend}}{360}$$

Radius to centre of metal is $0.125 + 0.032$

$$\text{B.A.} = 2 \times 3.1416 \times 0.157 \times \frac{90}{360} = 0.246.$$

The development would be as shown in Fig. 3c.

In aircraft work it is not usual for the draughtsman to state the required inside radius of a bend, but it is usual to make the inside radius twice the thickness of the metal, and this inside radius can be taken as correct for all sheet metal fitting bends, and agrees with Air Ministry demands.

The development of a similar angle plate to that of Fig. 3A but with the inside radius twice the gauge (2G), would be as shown in Fig. 4.

M would be $1 \text{ in.} - 2G + 1G$ (G = 0.064)

$$= 1 \text{ in.} - 3G$$

$$= 1 - 0.192 \text{ in.}$$

$$= 0.808 \text{ in.}$$

N would be $1.5 - 0.192 = 1.308.$

The bending allowance B.A. would be—

$$2 \times 3.1416 \times \text{radius to centre of metal} \times \frac{90}{360}$$

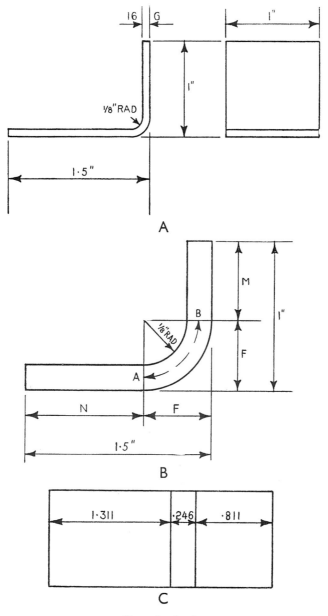

Figs. 3a, 3b, 3c

(Radius to centre $= 2\frac{1}{2}$G)

$$= 2 \times 3\cdot 1416 \times 0\cdot 16 \times \frac{90}{360} = 0\cdot 251.$$

1·308	·251	·808

FIG. 4

"Sight" Lines for Bending Purposes

Referring to Fig. 3A, simple angle plate, the next difficulty is to bend it correctly.

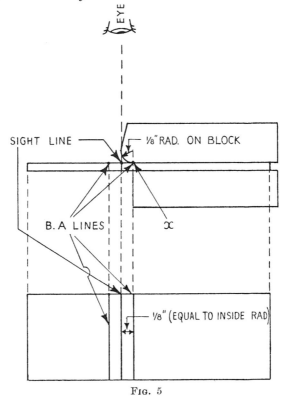

FIG. 5

This difficulty is overcome by the addition of a "sight" line. The plate must be bent around a bending block which has the correct radius on its edge; in this case the radius on the block would be $\frac{1}{8}$ rad. (Fig. 5). The correct position of the plate is shown in Fig. 5.

When in position in bending block one bending allowance line must be directly under the radius point of the block. This position is obtained by scribing a line (the sight line) at a distance equal to the inside radius of bend, from the bending allowance line (x in sketch) nearest the side being clamped in the block.

This line will then come directly in line with the edge of the block when the plate is in position, and can easily be "sighted" with the eye directly over the edge of the block.

Developments of More Difficult Fittings having Right-angled Bends

Ex. 1. The methods employed in the development of the simple angle plate can be applied to any fitting, simple or complex, which has right-angled bends.

For example refer to the U bracket shown in Fig. 6A.

The inside radius is not stated so it is reckoned to be 2G.

Separate bends from flats. (Fig. 6B.)

A (flat) $= 1$ in. $- 3G = 1 - 0 \cdot 144 = 0 \cdot 856$ in.

C (flat) $= 1$ in. $- 3G = 1 - 0 \cdot 144 = 0 \cdot 856$ in.

$$B.A. = 0 \cdot 188 \text{ in.}$$

B (flat) $= 2$ in. $- 6G$ (3G each side)

$$= 2 \text{ in.} - 0 \cdot 288 = 1 \cdot 712 \text{ in.}$$

The development is shown in Fig. 6C.

Ex. 2. See Fig. 7A.

The development of this type of fitting is shown in Fig. 7B and is practically self-explained.

The only difficulty may be to ascertain to position of the edges of the flanges at x and y, but, after studying Fig. 7A, it will be easily seen that these edges are (as marked) 2G and 3G respectively from the continuation of the inner bend allowance lines.

Ex. 3. See Fig. 8A.

This development is perfectly straightforward, but it will be

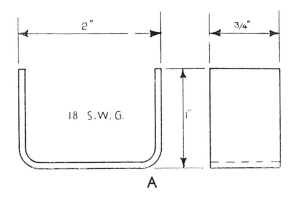

A

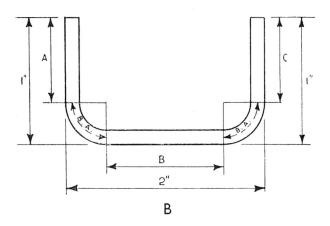

B

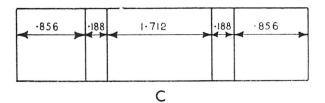

C

Figs. 6a, 6b, 6c

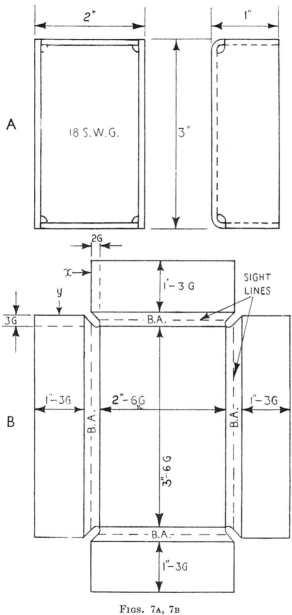

FIGS. 7A, 7B

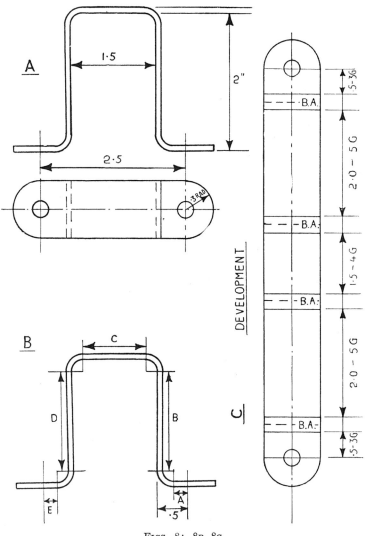

FIGS. 8A, 8B, 8C

noted that some measurements are given to the inside of the
metal and some to the outside.

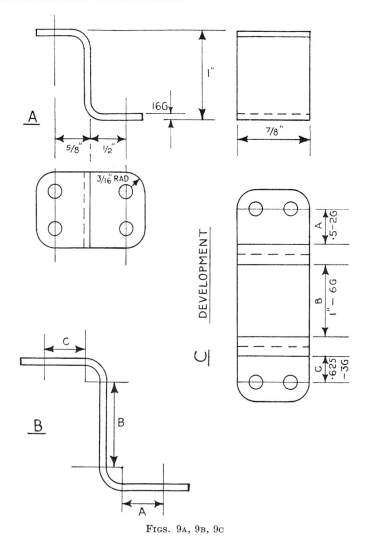

FIGS. 9A, 9B, 9C

It is only necessary to subtract 2G from inside measurements when calculating lengths of flats.

Flat portion C will therefore be 1·5 in. — 4G (2G each side).

Flat portions B and D will be 2 in. — 5G (2G top side and 3G at bottom).

Ex. 4. See Figs. 9A and 9B.

Ex. 5. See Figs. 10A and 10B.

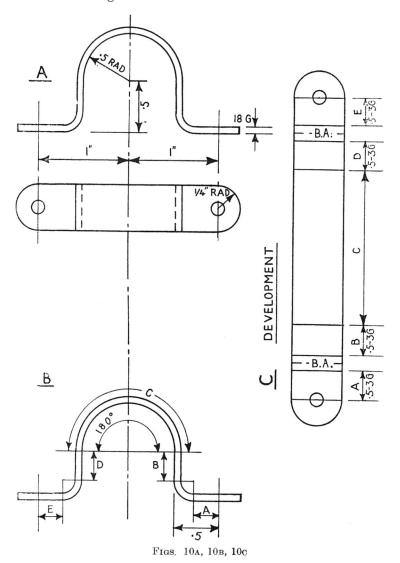

FIGS. 10A, 10B, 10C

Dimension C is developed in the same way as a small bending allowance.

$$C = 2\pi r \text{ (to centre of metal)} \times \frac{\text{angle of bend}}{360}$$

$$= 2 \times 3\cdot1416 \times 0\cdot524 \times \frac{180}{360} = 1\cdot646.$$

$$C = 1\cdot646.$$

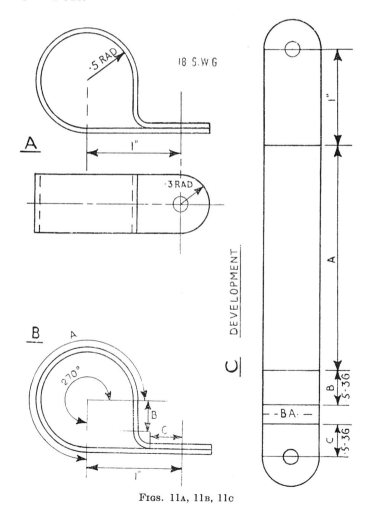

FIGS. 11A, 11B, 11C

Ex. 6. See Figs. 11A and 11B.

$$\text{Dimension } A = 2 \times 3{\cdot}1416 \times 0{\cdot}524 \times \frac{270}{360} = 2{\cdot}47 \text{ in.}$$

Ex. 7. Other examples of right-angle bend developments are shown in Figs. 12A to 13C.

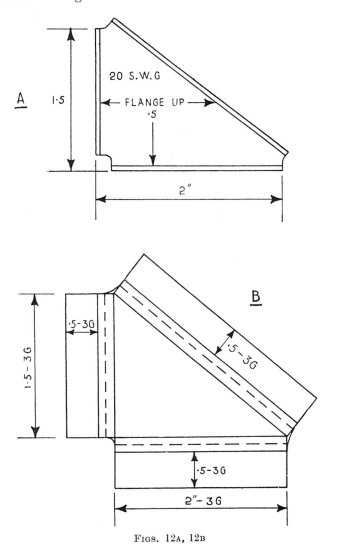

FIGS. 12A, 12B

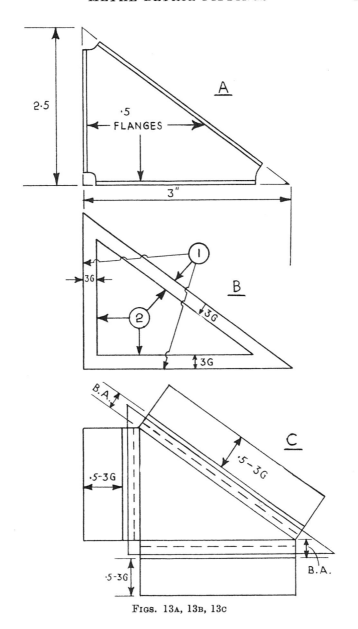

Figs. 13a, 13b, 13c

When dimensions to apex points are given, such as those in Fig. 13A, proceed as follows—

1. (Fig. 13B.) Scribe out profile of finished part.

2. Scribe three more lines at a distance 3G inwards from each profile line.

These second three lines will then become the beginnings of the three bending allowances for the flanges.

3. Proceed as shown in Fig. 13c, which is self-explanatory.

The methods employed in these examples should enable the fitter to develop any sheet metal fitting which has right-angled bends only.

For the accurate development of fittings having bends at angles either smaller or greater than a right angle, an elementary knowledge of trigonometry is necessary. This is dealt with in Chapter II.

CHAPTER II

TRIGONOMETRY

ALL technical details, proofs, etc., which would prove laborious have been left out, and calculations have been kept as easy as possible.

Trigonometry, as far as we are concerned, means calculations concerning right-angled triangles.

In a right-angled triangle ABC with angle B the right angle (Fig. 14)—

The side AC is known as the hypotenuse and is always the longest side and is opposite the right angle. The side AB is said to be adjacent to angle A or opposite to angle C. The side BC is said to be opposite angle A or adjacent to angle C.

Adjacent means "Adjoining" or "Next to."

Sines, Cosines, and Tangents

The sine of an angle means the proportion of the side opposite to the hypotenuse, or expressed as a simple equation

$$\text{Sine } A = \frac{\text{side opposite to } A}{\text{hypotenuse}}$$

The cosine of an angle means the proportion of the side adjacent to the hypotenuse

$$\text{Cosine } A = \frac{\text{side adjacent to } A}{\text{hypotenuse}}$$

The tangent of an angle is the proportion of the side opposite to the side adjacent

$$\text{Tangent } A = \frac{\text{side opposite}}{\text{side adjacent}}$$

In a triangle of any size, if the angle A remains unchanged the proportions of the sides to one another are always the same.

17

For instance, if the angle A is 30°, the sine of 30° or $\dfrac{\text{side opposite}}{\text{hypotenuse}}$ is always the same.

If, by careful measurement BC (side opposite) was found to be 1 in., and the hypotenuse AC was found to be 2 in., we could state that

$$\text{Sine } 30° = \frac{\text{side opposite}}{\text{hypotenuse}} = \tfrac{1}{2} = 0.5.$$

This would be true of sine 30° no matter what size the triangle was.

All of the values of the sines, cosines and tangents of all

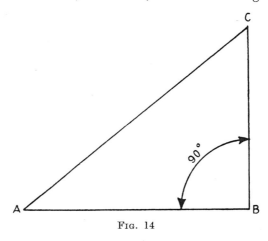

Fig. 14

angles have been worked out and are given in Table I on page 44; and it is from these values that all calculations are made.

For the purpose of developing, it is sufficiently accurate to work to the nearest half degree.

In the solution of triangles there is always a certain amount of information known of the triangle, with the necessity of finding unknown sides or angles. This is done by rearranging the equations on the previous page, putting in all known quantities, and calculating the unknown by ordinary arithmetic. This may prove, to some, very puzzling, so all these equations have been rearranged in Table II on page 46 to cover every calculation that will be necessary.

Ex. 1.   Given angle $A = 30°$

$$BC = 1·5 \text{ in.}$$

Required to find AB.

We have angle and opposite side given, to find adjacent side.

Refer to the table.

We are given "angle and opposite." Look along the top of the table, under the heading "When Given," until "Angle and Opposite" is found. Look down the left-hand column under the heading "To find" until you come to "Adjacent."

Follow the "Adjacent" column across, and follow the "Angle and Opposite" column down, until the two meet. At this point will be found the necessary information,

i.e. $\text{Adjacent} = \dfrac{\text{opposite}}{\text{Tan. of angle}} = \dfrac{1·5}{\text{Tan. } 30°}$

$$AB = \frac{1·5}{0·5774} = 2·598 \text{ in.}$$

Ex. 2.   Given angle $A = 40°$

Opposite side $BC = 2$ in.

To find side AC (hypotenuse).

Refer to table, which says "Given angle and opposite; To find hypotenuse."

$\text{Hypotenuse} = \dfrac{\text{opposite}}{\text{Sine of angle}} = \dfrac{2 \text{ in.}}{\text{Sine of } 40°}$

$$AC = \frac{2}{0·6428} = 3·111 \text{ in.}$$

Ex. 3.   Given angle $C = 46°$

Side $BC = 1$ in.

To find hypotenuse AC.

We are "given" angle C and side *adjacent* to C.

"To find" hypotenuse.

$\text{The table says hypotenuse} = \dfrac{\text{adjacent}}{\text{Cosine of angle}} = \dfrac{1 \text{ in.}}{0·6947}$

$$AC = 1·439.$$

Ex. 4.   Given $AB = 2$ in.

$$AC = 3 \text{ in.}$$

To find angle A.

"Given" adjacent and hypotenuse. "To find" angle.

Table says Cosine of angle $= \dfrac{\text{adjacent}}{\text{hypotenuse}} = \dfrac{2}{3} = 0\cdot6666.$

Cosine $A = 0\cdot6666.$

Note here that we have now found the *Cosine of the Angle,* NOT the angle.

The angle itself is found from the tables of Cosines, by finding the angle (to the nearest half degree) of which $0\cdot6666$ is the cosine.

The nearest angle which has a cosine value $0\cdot6666$ is $48°$, therefore angle $A = 48°$.

Ex. 5.   Given $AB = 4$ (opposite)

$$BC = 5 \text{ (adjacent).}$$

To find angle C.

We are "given" opposite and adjacent. "To find" angle.

The table says Tangent of angle $= \dfrac{\text{opposite}}{\text{adjacent}} = \dfrac{4}{5} = 0\cdot8$

Tan. $C = 0\cdot8$

From table of tangents we then find: $C = 38\tfrac{1}{2}°$.

Square Roots

In the triangle ABC.

The length of the hypotenuse squared equals the lengths squared respectively of the other two sides,

or $AC^2 = AB^2 + BC^2.$

It can be easily seen from the above equation that

$$AB^2 = AC^2 - BC^2$$

and $BC^2 = AC^2 - AB^2.$

If any two are given it is possible to find the third.

Ex. 1.   Given $AB = 3$ in.

$BC = 4$ in.

to find AC.

From the equations we get

$$AC^2 = AB^2 + BC^2$$
$$= 3^2 + 4^2$$
$$= 9 + 16$$
$$= 25.$$

Therefore $AC = \sqrt{25}$

[The symbol $\sqrt{}$ means "square root of," i.e. $\sqrt{9} = 3$.]

$$= 5 \text{ in.}$$

To Find the Square Root of a Number

The method for finding the square root of a number is dealt with in all elementary schools, but in case some may have forgotten, an example is set out and explained.

To find the square root of 18·25.

It is sufficiently accurate to work to three places of decimal.

```
        4· 2  7  2
     4| 18·25,00,00
        16
    82| .225
        .164
   847| ..6100
        ..5929
  8540| ...17100
```

(1) Add sufficient noughts to number to make six places of decimal.

(2) From the decimal point, mark off the figures in pairs each side.

(3) Take the first pair of figures and ascertain the largest number which, multiplied by itself, equals or is less than these first two figures.

4 is the largest number. Put down 4 on the left-hand side of the 18 and also in the quotient above the 18. $4 \times 4 = 16$. Put 16 below the 18 and subtract, leaving 2.

(4) Bring down the next pair of figures, i.e. 25.

(5) A figure must be added to the left of the 225· as the next divisor. Double the number now in the quotient (4) making 8, and enter this number in line with the 225.

(6) A figure must now be added to the 8 so that the complete number formed is the next divisor. This number to be added must be the next number for the quotient, that is, if a 2 is added making 82 the number 82 must be multiplied by the 2 making 164, which is to be put under the 225 as in long

division, followed by ordinary subtraction. The number added to the 8 must be the largest possible without exceeding the 225 after multiplication.

(7) Double the number which is now in the quotient—that is, 84—and put it down to the left of the 61 remainder.

(8) Bring down the next pair of figures, that is 00, and continue as before, adding figure 7 to the 84 and so on. The result to three places of decimal is 4·272. Each number in the quotient comes directly above two figures in the divisor.

Here are two more examples without explanation—

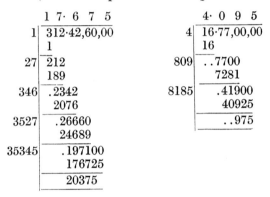

Ex. 2. In a right-angled triangle ABC, given $AC = 0·33$, $AB = 0·21$, find BC.

$$BC^2 = AC^2 - AB^2$$
$$= 0·33^2 - 0·21^2$$
$$= 0·1089 - 0·0441$$
$$= 0·0648$$

Therefore, $BC = \sqrt{0·0648} = 0·254$.

Any third side can be found from the other two in this manner.

It should now be possible for the fitter to solve any problem contained in a right-angled triangle.

Application of Trigonometry to Development

All examples are shown in the straight strip form, for simplicity.

FIGURE 15. The most common form of the application of trigonometry for development is that of the tube clip shown in Fig. 15A.

In Fig. 15A the small bends (2G inside rad.) run into the large bend (0·5 inside rad.).

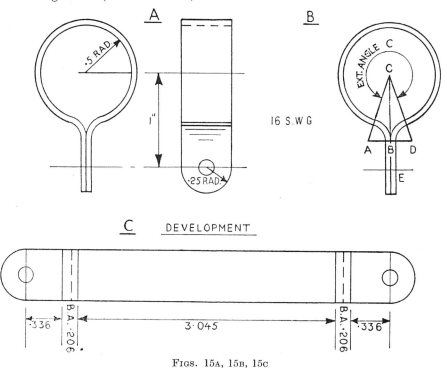

Figs. 15A, 15B, 15C

These bends are separated by joining the radius points *A* and *C*.

Join *A* and *D* thereby separating the small bends from the flats.

It will be seen that the development will consist of—

The two flat portions *BE*.

The two small B.A.'s contained by angles at *A* and *D*, and the large B.A. contained by the exterior angle at *C*.

Procedure

1. The length of *BE* is found by subtracting the length *BC* from the given dimension 1 in.

BC is "found" in the triangle *ABC*.

2. The bending allowances at *A* and *D* which are equal are calculated as follows—

$$\text{B.A.} = 2 \times \pi \times r \text{ (to centre of metal)} \times \frac{\text{angle of bend}}{360} \text{ } (A \text{ or } D)$$

The angle *A* can be found in the triangle *ABC*.

3. The large bending allowance contained by the exterior angle *C* is calculated as follows—

$$\text{B.A.} = 2 \times \pi \times r \text{ (to centre of metal)}$$
$$\times \frac{\text{angle of bend (Ext. } C)}{360}$$
$$= 2 \times 3\cdot1416 \times 0\cdot532 \times \frac{\text{Ext. } C \text{ angle}}{360}$$

The Exterior angle at *C* is found by subtracting angle *ACD* from 360°.

The angle *ACD* equals twice angle *ACB*.

Angle *ACB* = 90 − angle *CAB* (which is already found).

In the triangle *ABC*, Fig. 15A: To find *BC*—

$$BC^2 = AC^2 - AB^2 = (0\cdot5 + 3\text{G})^2 - (3\text{G})^2 = 0\cdot692^2 - 192^2$$
$$= 0\cdot478 - 0\cdot036 = 0\cdot442.$$

Therefore $BC = \sqrt{0\cdot442} = 0\cdot664.$

To find angle *BAC*.

In the table we want "to find" angle "when given" adjacent and hypotenuse (*AB* and *AC*).

$$\text{The table says cosine of angle} = \frac{\text{adjacent}}{\text{hypotenuse}} = \frac{AB}{AC} = \frac{0\cdot192}{0\cdot692}$$

Cosine *BAC* = 0·277

From cosine tables angle *BAC* = 74°

Angle *ACB* = 90° − *BAC*
= 90° − 74°
= 16°

Exterior angle *C* = 360 − angle *ACD*. *ACD* = twice *ACB*
= 360° − 32°   = 32°
= 328°

1. $BE = 1$ in. $- BC$ $BC = 0.664$

$$= 1 \text{ in.} - 0.664 = 0.336.$$

2 Small B.A.s at A and D

$$= 2 \times 3.1416 \times \text{radius (to middle of metal)} \times \frac{BAC}{360}$$

radius $= 2\tfrac{1}{2}G = 0.16.$ $BAC = 74°.$

Therefore B.A. $= 2 \times 3.1416 \times 0.16 \times \dfrac{74}{360} = 0.206.$

3. The B.A. contained by exterior angle at C (328°)

$$\text{B.A.} = 2 \times 3.1416 \times 0.532 \times \frac{328}{360} = 3.045.$$

The development of this clip is shown in Fig. 15c. No matter how large or small the bending allowance is, the sight line comes at a distance equal to the inside radius from the B.A. line on the side which is to be gripped in the blocks.

Ordinarily for this clip the lugs would be bent first by gripping the centre portion of the clip in the bending blocks. The position of the sight lines is shown in Fig. 15c.

FIGURE 16. Fig. 16A gives another typical fitting which must be developed by trigonometry (no holes are put in this fitting to make it more simple).

In this sketch measurements are given to apex point, which is usual for this type of fitting.

As in all developments separate all bends from flats, that is, by dropping perpendiculars from F on to the flats; the perpendiculars being FD and FE (see Fig. 16B).

The development consists of EA (flat) DC (flat), and the B.A. contained by angle DFE.

The flat $EA = AB - BE$

$$= 1 \text{ in.} - BE$$

BE is not known.

To find BE: Join BF.

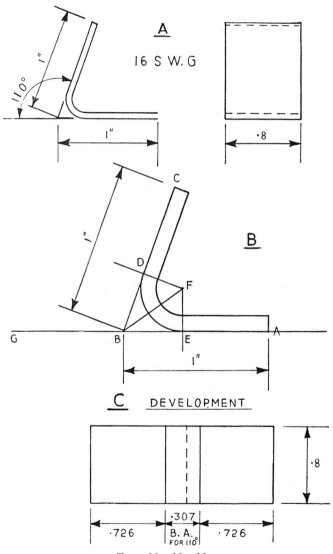

A

16 S W. G

B

C   DEVELOPMENT

·726   B. A.
FOR 110°   ·726

·307

·8

FIGS. 16A, 16B, 16C

Then *BE* can be found in the triangle *BEF*, but first of all it is necessary to see what is already known of triangle *BEF*.

$$EF = 3G$$

Angle *EBF* = half of angle *EBD* (*EBD* = 180° − 110° = 70°),

$$= \frac{70}{2} = 35°.$$

We have now "given" angle *EBF* (35°) and "opposite" *EF* (3G).

To find adjacent—

The table says adjacent

$$= \frac{\text{opposite}}{\text{Tan. of angle}} = \frac{0\cdot192}{\text{Tan. } 35°} = \frac{0\cdot192}{0\cdot7} = 0\cdot274$$

The flat *EA* = 1 in. − *EB* = 1 in. − 0·274 = 0·726 in.

,,   *DC* = *EA* = 0·726 in.

The bending allowance angle is *DFE* which is equal to the outside angle *GBC* = 110° (given). (The bending allowance angle is equal to the outside angle as shown in Fig. 16D.)

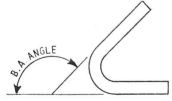 

FIG. 16D

The development of Fig. 16A is shown in Fig. 16C.

FIGURE 17. The typical examples shown in Figs. 17 A, B, C, to Figs. 22 A, B, C, show the construction lines and triangles necessary for their development, and a careful study of these developments will enable the fitter to develop any fitting which will be encountered.

Construction lines are shown on Fig. 17B.

The development shown on Fig. 17C consists of Flats *KD*, GF and *JH*.

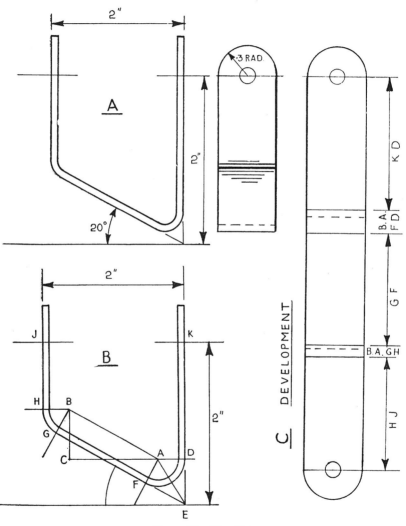

Figs. 17a, 17b, 17c

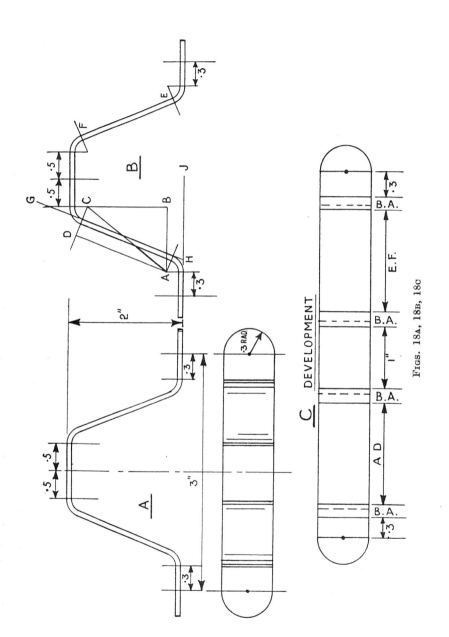

Figs. 18a, 18b, 18c

Two B.A.'s contained by angles *HBG* and *FAD*.

1. $KD = 2$ in. $- DE$. *DE* is "found" in triangle *AED*.
2. $GF = AB$. *AB* is "found" in triangle *ABC*.
3. $JH = 2$ in. $- (DE + BC)$. *DE* is already found, *BC* is found in triangle *ABC*.
4. B.A. Contained by angle *FAD* is worked out as usual, the angle of the bend being $90° + 20° = 110°$.
5. B.A. Contained by angle *GBH*; angle of bend is 70°.

FIGURE 18. In this type of fitting the flats at the base and at the top are given by the draughtsman. Fig. 18B shows construction lines and Fig. 18c development.

It is only necessary to find the B.A. angles, and the flats *AD* and *EF* (which are equal).

Construct lines as shown.

1. In △ *ABC* find *AC* and angle *CAB*.
2. In △ *ACD* find *AD* and angle *DAC*.

The B.A. angles (all four being equal) are all equal to that contained by angle *GHJ* which equals angle *DAB*.

Angle $DAB = DAC + CAB$ (both of which have been found).

The flat $EF = AD$ (already found).

FIGURE 19A. This is rather a complicated development, so very careful study will be necessary.

FIGURE 19B shows construction lines and Fig. 19c development.

1. In the triangle *BCD*,
    Calculate angle *DCB*
    Length of *BD*.

2. In the triangle *BFM*,
    Calculate *MBF*
    Length of *MF*.

The development is composed of—

1. Two flats $DE = 1$ in. $- BD$.
2. B.A.'s contained by angles

$$BAD \text{ and } BCD = 2\pi r \text{ (centre of metal)} \times \frac{BCD}{360}$$

3. B.A.'s contained by angles

$$CBG \text{ and } ABK = 2\pi r \times \frac{CBG}{360}$$

where r = radius to centre of metal

$$= \tfrac{3}{8}\text{in.} + \tfrac{1}{2}G$$

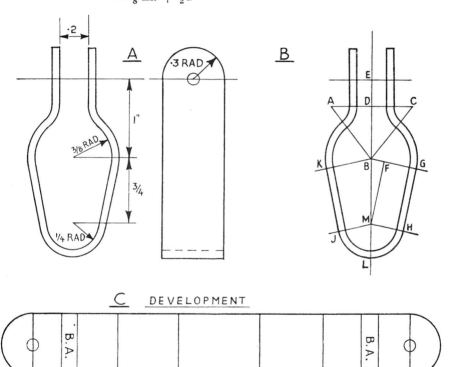

FIGS. 19A, 19B, 19C

4. The two flats GH and JK. $GH = MF$ (which has been found).

5. B.A. contained by angle JMH

Angle JMH = twice angle LMH. $LMH = MBF$ **(found)**
$$= \text{twice } MBF.$$

Development of a Hinge

Ex. (see Fig. 20A). The 2 in. measurement is given to the centre of the hinged portion.

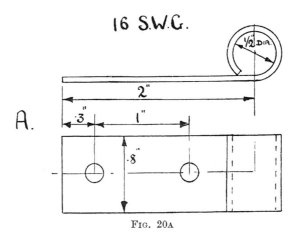

A.

FIG. 20A

The development consists of (see Fig. 20B)—
Flat portion *AB* which is 2 in., plus the bending allowance taken up by the hinge.

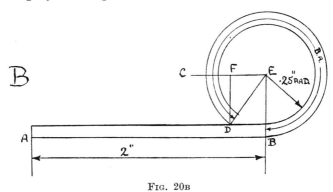

FIG. 20B

Construct line *EC* parallel to *AB* and drop perpendicular *DF* on to it from *D*; join *DE*.

The B.A. for the hinge is contained by the exterior angle *BED*.

Exterior angle

BED = exterior angle BEC + Angle DEC

exterior angle $BEC = 270°$ ($\frac{3}{4}$ of 360°)

Therefore exterior angle

$BED = 270°$ + angle DEC.

Angle DEC is calculated from the triangle DEF in which is known two sides.

$$DF = 0\cdot25$$
$$DE = 0\cdot25 + 1G.$$

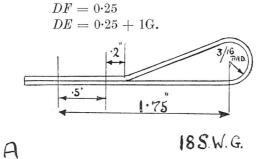

A

18 S.W.G.

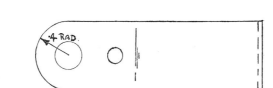

Fig. 21A

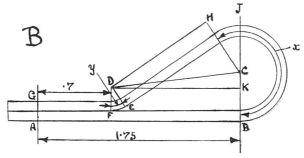

B

Fig. 21B

FIGURE 21A. The development of this fitting (see Fig. 21B) consists of—

AB (1·75 in.), bending allowance x, DH (Flat), bending allowance y, and GF (0·7 in).

Portion *DH* and bending allowance *y* are calculated in the same way as for development of Fig. 18c.

Bending allowance *x* is contained by exterior angle *BCH* which is 180° plus angle *HCJ*.

Angle *HCJ* equals angle *EDF* which has already been found for calculating bending allowance *y*.

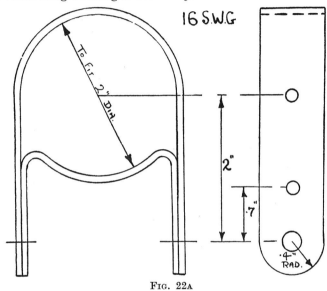

16 S.W.G

To fit 2" Dia.

2"

·7"

·4" RAD.

FIG. 22A

FIGURE 22A. This bracket consists of two parts.

The development of the outer part is simple and is composed of two flat portions 2 in. each and a B.A. for 180° (1 in. rad.).

The development of the inner portion, however, is more difficult. It is composed of (see Fig. 22B)—

Flats *AB* and *GH* (equal), two B.A.'s *x* and *y* (equal), and one B.A. *z* (1 in. rad.).

$$AB \text{ or } GH = 2 \text{ in.} - EF.$$

EF is calculated from the triangle *DEF* in which

$$DE = 1 \text{ in.} - 3G$$
$$FD = 1 \text{ in.} + 3G$$

B.A. *y* is contained by angle *GDF*.

$$\text{Angle } GDF = 180° - \text{angle } EDF.$$

Angle *EDF* is calculated from the triangle *DEF*.

B.A. z is contained by the angle *CFD*.

> Angle *CFD* = twice angle *DFE*
> Angle *DFE* = 90° − *EDF* (already found).

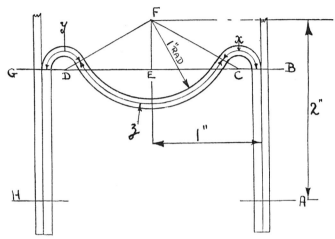

Fɪɢ. 22ʙ

The construction lines necessary in the past examples of development must be memorized, but it becomes obvious where these lines must come when a fitter has had considerable practice in this work.

These examples cover practically every type of development, but perhaps in a different form; all the principles, however, are applied in exactly the same way.

Development by Projection. See Fig. 23ᴀ.

To develop this fitting proceed as follows—

1. Construct to scale on metal the centre line of the cross-section (the bends can be omitted).

2. Construct the other view dead in line as shown in Fig. 23ʙ.

3. On (1) divide flats from bends at *A* to *K*.

4. Project all points *A* to *K* across parallel to the second view as shown in sketch.

5. Scribe perpendicular *MN*.

6. Work out developed lengths and lay out on metal with bend lines in position, as shown in sketch. (Fig. 23c.)

7. Scribe perpendicular across all lines from *M* to *N*.

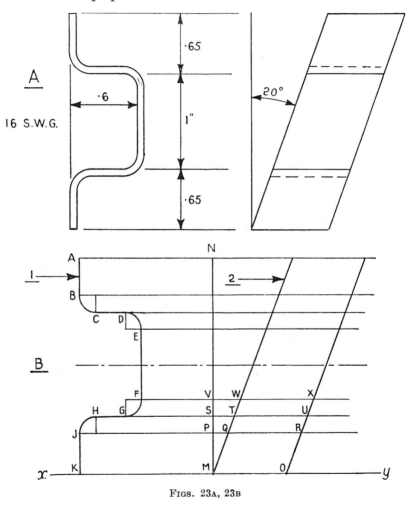

Figs. 23A, 23B

8. (*a*) Set dividers at distance equal to *MO* (Fig. 23B), and transfer this measurement on to the appropriate line on the development (line *K*). (*b*) Set dividers to *PQ* (Fig. 23B) and transfer to line *J*. Then transfer *PR*.

9. Points *H* and *G* (Fig. 23B) are projected across on the same line so that on the development we get the same transferred measurements *ST* and *SU* on the two lines *H* and *G*.

10. Continue transferring the measurements on all lines *A* to *K*.

11. Join the points *M.Q.T.T.*, etc., and *O.R.U.U.*, etc., and the development will be ready for the "sight" lines.

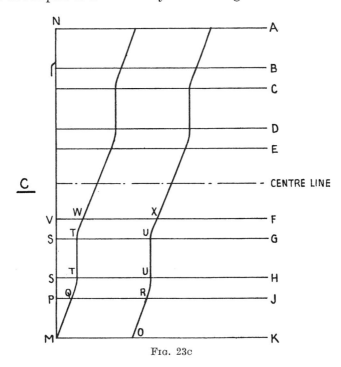

FIG. 23c

Fig. 24A. Proceed as in (*b*) and (*c*) as before.

3. Divide centre line into any number of equal parts, about six will do for job of this size.

4. Project across points *A* to *G* across as before.

5. Scribe two lines the developed length apart and divide the length into six equal parts, scribe perpendicular *AG*.

6. Proceed transferring measurements as before, and finally join up the points.

The development of Fig. 25A is carried out in a similar

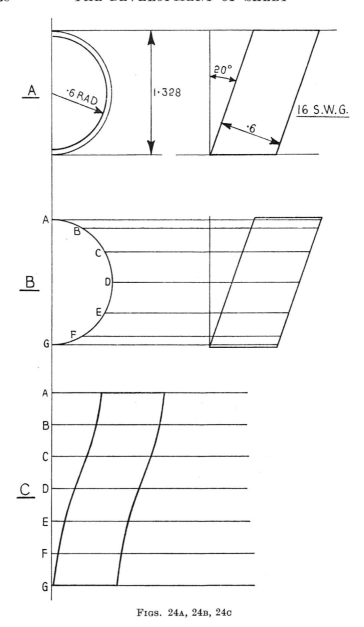

FIGS. 24A, 24B, 24C

manner as the previous two, and the procedure is clearly shown in the sketches.

To Develop a Cone Frustrum Surface

See Fig. 26A.

Operations

1. Construct, on metal to be used, the elevation *BCJH* as shown in Fig. 26B.

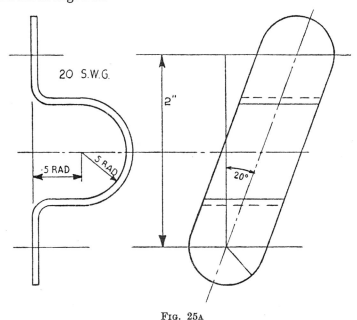

Fɪɢ. 25ᴀ

2. Continue the centre lines of the thicknesses to the point of intersection *A*.

3. Set dividers at *AB* and scribe an arc *BD*.

4. Set dividers at *AC* and scribe an arc *CE*.

5. The development, shown shaded, is contained by a certain angle *CAE* which must be calculated as follows—

The length of arc *CE* is equal to the circumference at *D2* which is ($\pi \times D2$).

The arc *CE* is the same proportion of a complete circumference (with *AC* as radius) as the angle *CAE* is of 360°.

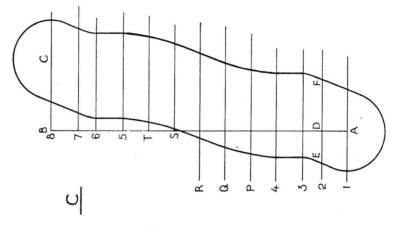

C|

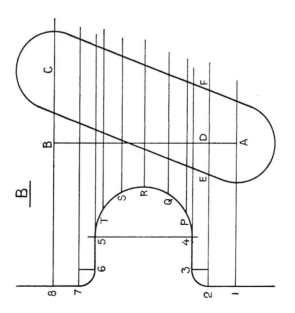

B

Figs. 25b, 25c

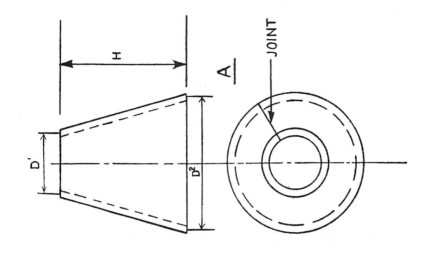

A

JOINT

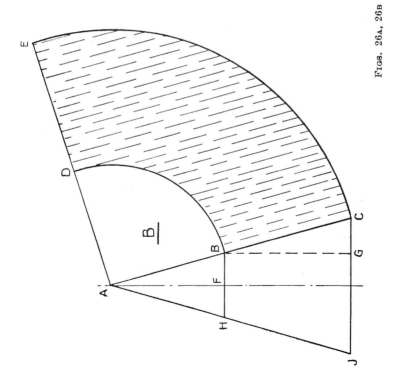

B

Figs. 26a, 26b

Circumference with AC as radius would be $2 \times \pi \times AC$. Therefore, stated as an equation

$$\frac{CAE}{360°} = \frac{\pi \times D2}{2 \times \pi} \times AC$$

The π's cancel and it follows that

$$\text{Angle } CAE = \frac{360 \times D2}{2 \times AC}$$

$$= \frac{180 \times D2}{AC}$$

As an alternative method, instead of constructing the elevation, the radii AB and AC can be calculated from the triangles BCG and ABF.

Typical Application of Development

FIGURE 27. Fitting "A" is developed using methods previously dealt with (page 26).

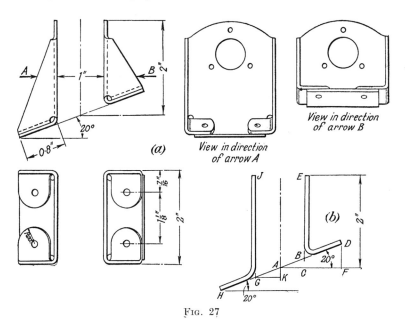

FIG. 27

However, the length BD (Fig. 27 (b)) is given ($\cdot 8$ in.), but BE must be calculated—

$$BE = CE - BC$$

$$= 2 \text{ in.} - BC$$

BC is found in triangle ABC. Given $AC = \frac{1}{2}$ in.; angle $CAB = 20°$.

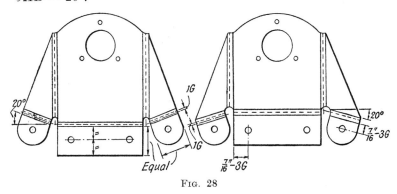

FIG. 28

In fitting "B"

AG is given ($\cdot 8$ in.)

$GJ = EC + BC \ (BC = AK)$

The method of laying out the developments, complete with lugs and webs, is clearly shown in the sketch (Fig. 28).

The bends on the lugs and webs will be seen to be ordinary right-angled bends.

TABLE I

De-grees	Sines	Cosines	Tan-gents	De-grees	Sines	Cosines	Tan-gents
0	0000	1·0000	0000	0·5	0087	1·000	0087
1	0175	9998	0175	1·5	0262	9997	0262
2	0349	9994	0349	2·5	0436	9990	0437
3	0523	9986	0524	3·5	0610	9981	0612
4	0698	9976	0699	4·5	0785	9969	0787
5	0872	9962	0875	5·5	0958	9954	0963
6	1045	9945	1051	6·5	1132	9936	1139
7	1219	9925	1228	7·5	1305	9914	1317
8	1392	9903	1405	8·5	1478	9890	1495
9	1564	9877	1584	9·5	1650	9863	1673
10	1736	9848	1763	10·5	1822	9833	1853
11	1908	9816	1944	11·5	1994	9799	2035
12	2079	9781	2126	12·5	2164	9763	2217
13	2250	9744	2309	13·5	2334	9724	2401
14	2419	9703	2493	14·5	2504	9681	2586
15	2588	9659	2679	15·5	2672	9636	2773
16	2756	9613	2867	16·5	2840	9588	2962
17	2924	9563	3057	17·5	3007	9537	3153
18	3090	9511	3249	18·5	3173	9483	3346
19	3256	9455	3443	19·5	3338	9426	3541
20	3420	9397	3640	20·5	3502	9367	3739
21	3584	9336	3839	21·5	3665	9304	3939
22	3746	9272	4040	22·5	3827	9239	4142
23	3907	9205	4245	23·5	3987	9171	4348
24	4067	9135	4452	24·5	4147	9100	4557
25	4226	9063	4663	25·5	4305	9026	4770
26	4384	8988	4877	26·5	4462	8949	4986
27	4540	8910	5095	27·5	4617	8870	5206
28	4695	8829	5317	28·5	4772	8788	5430
29	4848	8746	5543	29·5	4924	8704	5658
30	5000	8660	5774	30·5	5075	8616	5890
31	5150	8572	6009	31·5	5225	8526	6128
32	5299	8480	6249	32·5	5373	8434	6371
33	5446	8387	6494	33·5	5519	8339	6619
34	5592	8290	6745	34·5	5664	8241	6873
35	5736	8192	7002	35·5	5807	8141	7133
36	5878	8090	7265	36·5	5948	8039	7400
37	6018	7986	7536	37·5	6088	7934	7673
38	6157	7880	7813	38·5	6225	7826	7954
39	6293	7771	8098	39·5	6361	7716	8243
40	6428	7660	8391	40·5	6494	7604	8541
41	6561	7547	8693	41·5	6626	7490	8847
42	6691	7431	9004	42·5	6756	7373	9163
43	6820	7314	9325	43·5	6884	7254	9490
44	6947	7193	9657	44·5	7009	7133	9827
45	7071	7071	1·0000	45·5	7133	7009	1·0176

TABLE I (*Contd.*)

De-grees	Sines	Cosines	Tan-gents	De-grees	Sines	Cosines	Tan-gents
46	7193	6947	1·0355	46·5	7254	6884	1·0538
47	7314	6820	1·0724	47·5	7373	6756	1·0913·
48	7431	6691	1·1106	48·5	7490	6626	1·1303
49	7547	6561	1·1504	49·5	7604	6494	1·1708
50	7660	6428	1·1918	50·5	7716	6361	1·2131
51	7771	6293	1·2349	51·5	7826	6225	1·2572
52	7880	6157	1·2799	52·5	7934	6088	1·3032
53	7986	6018	1·3270	53·5	8039	5948	1·3514
54	8090	5878	1·3764	54·5	8141	5807	1·4019
55	8192	5736	1·4281	55·5	8241	5664	1·4550
56	8290	5592	1·4826	56·5	8339	5519	1·5108
57	8387	5446	1·5399	57·5	8434	5373	1·5697
58	8480	5299	1·6003	58·5	8526	5225	1·6319
59	8572	5150	1·6643	59·5	8616	5075	1·6977
60	8660	5000	1·7321	60·5	8704	4924	1·7675
61	8746	4848	1·8040	61·5	8788	4772	1·8418
62	8829	4695	1·8807	62·5	8870	4617	1·9210
63	8910	4540	1·9626	63·5	8949	4462	2·0057
64	8988	4384	2·0503	64·5	9026	4305	2·0965
65	9063	4226	2·1445	65·5	9100	4147	2·1943
66	9135	4067	2·2460	66·5	9171	3987	2·2998
67	9205	3907	2·3559	67·5	9239	3827	2·4142
68	9272	3746	2·4751	68·5	9304	3665	2·5386
69	9336	3584	2·6051	69·5	9367	3502	2·6746
70	9397	3420	2·7475	70·5	9426	3338	2·8239
71	9455	3256	2·9042	71·5	9483	3173	2·9887
72	9511	3090	3·0777	72·5	9537	3007	3·1716
73	9563	2924	3·2709	73·5	9688	2840	3·3759
74	9613	2756	3·4874	74·5	9636	2672	3·6059
75	9659	2588	3·7321	75·5	9681	2504	3·8667
76	9703	2419	4·0108	76·5	9724	2334	4·1653
77	9744	2250	4·3315	77·5	9763	2164	4·5107
78	9781	2079	4·7046	78·5	9799	1994	4·9152
79	9816	1908	5·1446	79·5	9833	1822	5·3955
80	9848	1736	5·6713	80·5	9863	1650	5·9758
81	9877	1564	6·3138	81·5	9890	1478	6·6912
82	9903	1392	7·1154	82·5	9914	1305	7·5958
83	9925	1219	8·1443	83·5	9936	1132	8·7769
84	9945	1045	9·5144	84·5	9954	0958	10·39
85	9962	0872	11·43	85·5	9969	0785	12·71
86	9976	0698	14·3	86·5	9981	0610	16·35
87	9986	0523	19·08	87·5	9990	0436	22·9
88	9994	0349	28·64	88·5	9997	0262	38·19
89	9998	0175	57·29	89·5	1·0000	0087	114·6
90	1·0000	0000					

TABLE II

WHEN GIVEN

To Find	Angle and Adjacent	Angle and Opposite	Angle and Hypotenuse	Adjacent and Opposite	Adjacent and Hypotenuse	Opposite and Hypotenuse
Angle				Tangent of Angle $= \dfrac{\text{Opp.}}{\text{Adj.}}$	Cosine of Angle $= \dfrac{\text{Adj.}}{\text{Hyp.}}$	Sine of Angle $= \dfrac{\text{Opp.}}{\text{Hyp.}}$
Opposite	Adj. × Tangent of Angle		Hyp. × Sine of Angle		$\sqrt{\text{Hyp}^2 - \text{Adj}^2}$	
Adjacent		$\dfrac{\text{Opp.}}{\text{Tangent of Angle}}$	Hyp. × Cosine of Angle			$\sqrt{\text{Hyp}^2 - \text{Opp}^2}$
Hypotenuse	$\dfrac{\text{Adj.}}{\text{Cosine of Angle}}$	$\dfrac{\text{Opp.}}{\text{Sine of Angle}}$		$\sqrt{\text{Adj}^2 + \text{Opp}^2}$		

TABLE III

TABLE OF BENDING ALLOWANCES

Inside Radius of Bends = 2G

Angle of Bend	Gauge (S.W.G.)					
	14G	16G	18G	20G	22G	24G
10°	·035	·028	·021	·016	·012	·010
20°	·070	·056	·042	·031	·024	·019
30°	·105	·084	·063	·047	·037	· ·029
40°	·140	·112	·084	·063	·049	·038
50°	·175	·140	·105	·079	·061	·048
60°	·209	·168	·126	·094	·073	·058
70°	·244	·195	·147	·110	·086	·067
80°	·279	·224	·168	·126	·098	·077
90°	·314	·251	·188	·141	·110	·086

INTERMEDIATE allowances are pro rata.

Use of Bending Allowance Tables

This table can only be used when the inside radius of the bend is the usual 2G.

Although angles in the table are at 10° intervals the tables cover every angle from 0° — 360°.

Ex. 1. B.A. required for 18 S.W.G. for 87°.

From the table

$$80° \text{ (18 S.W.G.)} = 0.168.$$

The remaining 7° is arrived at by dividing the B.A. for 70° by 10 or in other words by moving the decimal point one place to the left.

$$70° \text{ B.A.} = 0.147$$

$$\therefore \quad 7° \text{ B.A.} = 0.0147$$

B.A. for 87° = 0·168 + 0·0147 = 0·183 (to three places of dec.).

Ex. 2. B.A. required for 153° for 20 S.W.G.—

$$90° \text{ (20 S.W.G.)} = 0.141$$

$$60° \qquad\qquad = 0.094$$

$$\underline{3° \text{ (1/10 of 30°)} = 0.0047}$$

$$153° \qquad\qquad \overline{0.240} \quad \text{(to three places of dec.).}$$

TABLE IV

STANDARD WIRE GAUGES
(Inches)

Gauge	Inches	Gauge	Inches
7/0	·5	14	·080
6/0	·464	15	·072
5/0	·432	16	·064
4/0	·400	17	·056
3/0	·372	18	·048
2/0	·348	19	·040
0/0	·324	20	·036
1	·3	21	·032
2	·276	22	·028
3	·252	23	·024
4	·232	24	·022
5	·212	25	·020
6	·192	26	·018
7	·176	27	·0164
8	·160	28	·0148
9	·144	29	·0136
10	·128	30	·0124
11	·116	31	·0116
12	·104	32	·0108
13	·092	33	·010